YOUR KNOWLEDGE HAS VALUE

- We will publish your bachelor's and master's thesis, essays and papers

- Your own eBook and book - sold worldwide in all relevant shops

- Earn money with each sale

Upload your text at www.GRIN.com and publish for free

Matthias Pilecky

The Influence of 3-Hydroxybutyrate and Microcurrent Treatment on Cardiomyocytes during Simulated Hypertrophy

GRIN Verlag

Bibliografische Information der Deutschen Nationalbibliothek:

Die Deutsche Bibliothek verzeichnet diese Publikation in der Deutschen National-
bibliografie; detaillierte bibliografische Daten sind im Internet über http://dnb.d-
nb.de/ abrufbar.

Imprint:

Copyright © 2010 GRIN Verlag GmbH
Druck und Bindung: Books on Demand GmbH, Norderstedt Germany
ISBN: 978-3-640-86955-8

This book at GRIN:

http://www.grin.com/en/e-book/168718/the-influence-of-3-hydroxybutyrate-and-
microcurrent-treatment-on-cardiomyocytes

The Influence of 3-Hydroxybutyrate and Microcurrent Treatment on Cardiomyocytes during Simulated Hypertrophy

2. Bachelorarbeit

Bachelorstudiengang Molekulare Biotechnologie

Fachhochschule Campus Wien

Vorgelegt von:
Matthias Pilecky

Abgabetermin: 07.06.2010

Abstract

Chronic Heart Failure (CHF) is combined with various metabolic shifts. The continuous adrenergic stress results in a metabolic shift increasing glycolysis similar to a fetal metabolism. However also insulin resistance (IR) was reported triggering low glucose uptake and mitochondrial uncoupling and therefore reactive oxygen species (ROS) production reduce cardiac contractility. The adrenergic increased free fatty acids (FFA) are metabolized to ketone bodies (mainly β-hydroxybutyrate (OHB)) which are also energy stocks for brain and heart. Elevated blood ketone levels have been reported during CHF similar to diabetes. Clinically a correlation between ketone body blood level and severity of CHF has been discovered. However it remains unclear whether this is a result of metabolic changes or a compensatory mechanism. The aim of this study was to show, that rat cardiomyocytes (H9c2) treated with Phenylephrine (PE) for hypertrophy are slightly more susceptible to OHB at concentrations similar found in patients with CHF and that OHB reduces cell area significantly.

Zusammenfassung

Chronic Heart Failure (CHF) geht einher mit dauerhafter hyperadringerger Stimulation der Kardiomyozyten, welche - ähnlich dem fötalen Metabolismus - einen „metabolic shift" von Fettsäure- zu Glucosemetabolismus, bewirkt. Es wurde berichtet, dass verschiedene zusätzliche Veränderung, wie Insulinresistenz und damit verbundene geringere GLUT4 Level, oder mitochondrielles Uncoupling, und auch damit verbundene erhöhte ROS Produktion die Leistungsfähigkeit des Herzens herabsetzen. Durch die erhöhten freien Fettsäureplasmalevel (FFA) und die generellen Veränderungen der Muskelmetabolismen produziert die Leber vermehrt Ketonkörper (KB) (besonders β-Hydroxybutyrat (OHB)). Ketonkörper sind neben Fettsäuren und Glucose Energielieferanten für das Herz. Während CHF können die Ketonkörperblutlevel bis auf 10 mM ansteigen. Viele Forschungsgruppen spekulierten, dass Ketonkörper den Citratcyclus hemmen, jedoch zeigten Pelletier et al. dass OHB die Glucoseaufnahme durch Inhibierung des AMPK/p38 MAPK Signalpathways erreicht, dessen Aktivierung eine erhöhte Glucoseaufnahme bewirkt. In dieser Arbeit wurde gezeigt, dass Ratten-Kardiomyozytenzellen (H9c2) im durch Phenylephrin induzierten hypertrophen Zustand leicht erhöhtere Aktivität bei OHB zeigen als unbehandelte, sowie dass sich die Zellgröße bei OHB Behandlung deutlich verkleinert.

Register

Introduction

Theoretical Background

Chronic Heart Failure

Acute heart failure (AHF) and chronic heart failure (CHF) are the most death causing diseases in industrial countries. In complex cardiac diseases various factors such as cardiac overload, sympathetic tone, diet or inherited factors play an essential role and finally cause mechanical imbalance between cardiac output (CO) and systemic vascular resistance (SVR) resulting in decreasing blood pressure, cardiac shock and eventually death (1). Whereas in AHF a distinct event triggers cardiac failure within minutes, CHF develops multiple metabolic, protein expression and humeral alterations over time (2). The exact mechanisms leading to CHF have not been identified exactly yet although high blood pressure, resulting from arteriosclerotic events, is considered as one of the major causes. According to the Frank-Starling-mechanism, which describes stroke work as relation between stroke volume (SV), ventricle end systolic and end diastolic pressure, with increasing preload (filling pressure) the respective recruited SV decreases (1). This is on the one hand combined with an enhanced sympathetic tone, on the other hand mechanoreceptors (caveolae) in the membrane of cardiomyocytes are stimulated leading to an shift in Ca^{2+} concentrations (3) and thus triggering of a cascade via caveolin-3 and STAT3 resulting in hypertrophic growth (4) and alteration in extracellular matrix (ECM) and expression of matrix metallo proteinasen (MMP) (5). In animal models with artificial pressure overload the expression of myc significantly increases. These facts suggest that during hypertension the myocardial cell cycle machinery is activated and not leading to proliferation but to hypertrophy (6). To avoid this, the atrium releases atrial natriuretic peptite (ANP) which acts diuretic and thus volume unloading (7). In case of CHF cardiomyocytes enter a vicious circle suffering contractile failure, which leads to ventricle remodeling causing further myocyte energetic imbalance (8). The worse the cardiac state is the more ANP and BNP is released. The blood levels of ANP and BNP (brain natriuretic peptide; released by the ventricles) are of diagnostic value in CHF, (1) as they are significantly elevated (9).

Mitochondrial Alterations

Changes in cell signaling due to the mechanisms mentioned above have significant effects on the mitochondria. Increased activation of myc directly influences PPAR-γ coactivator-1α (PGC-1) (6) whose overexpression upregulates the mitochondrial transcription factor (mtTFA) leading to mitochondrial proliferation (10). An increased amount of mitochondria in hypertrophic cardiomyocytes is accompanied by a smaller volume and abnormal functionality, as described in a dog model (11). Although various reports on decreasing mitochondrial complex activity exist the exact mechanism remains unclear. A significant decrease was described for Complex I, III and IV in dog (12) and mice (13) models, assuming damaged mtDNA, since their subunits are mostly mitochondrial encoded and the expression of mitochondrial gene is regulated by mtDNA copy number (14). In contrast other clinical investigations only found a significant decrease in complex I assuming that the mtDNA is intact (15).

Metabolic Shift

The human heart in general is a metabolic omnivore (8). While the fetal heart mainly receives the bigger part of energy from glycolysis and lactate oxidation the growing heart prefers fatty acids as source of energy (16). In CHF the energy reserves of cardiomyocytes considerably decrease as seen in the ratio of Phosphocreatine to ATP (PCr/ATP) which also correlates with the severity of the disease (17). Under physiological conditions the heart adopts its metabolism in response to hormones, changes in workload or blood supply as well as altered gene expression (18). Basically a genetic metabolic shift away from using fatty acids to glucose as main energy source comparable to processes in fetal organism has been observed in CHF (19). This process seems to be a compensatory mechanism for managing increased workload. Studies performed in a rat model have shown an overexpression of insulin-independent glucose transporter GLUT-1. increased cardiac performance and prevented heart failure even in an artificial pressure-overload induced hypertrophic state without influencing glucose homeostasis (20). However in CHF increased sympathetic tone resulting in elevated noradrenalin levels constricts coronary vessels leading to minor oxygen supply and elevates FFA levels (21) (22). Enhanced FFA levels also influence mitochondrial uncoupling and uncoupling proteins (23) which are also a source of ROS (24).

Insulin Resistance

Elevated FFA levels as a result of the hyperadrenergic state promote an increasing insulin resistance (IR) and inhibit glucose uptake as well as glycolysis by heart and muscle cells (25), by blocking the signal cascade necessary for the fusion of insulin-dependent glucose transporter GLUT-4 vesicle with the plasma membrane (26). Accelerating IR may be the reason for increased FFA uptake compared to glucose shown by cardiomyocytes during HF (27) (28) since a decrease of FFA levels improves myocardial glycolysis (29). However several studies showed no difference in insulin or glucagon blood levels (9). In conclusion a metabolic vicious circle of elevated FFA causing IR, preventing efficient energy substrate usage (and causing ROS) leading to more adrenergic stimulation establishes additionally to the mechanical vicious circle previous mentioned (30).

Ketone Bodies

Chronic elevated FFA levels also alters liver metabolisms which start converting FFA into ketone bodies (KB) namely β-Hydroxybutyrat (OHB) and Acetoacetat (AAc) (31). Some papers referred to KB as "superfuel" since it produces more energy per oxygen consumption than glucose or fatty acids, because concentrations of NADH increases relative to NAD^+ and CoQ relative to $CoQH_2$ accelerating the redox-reaction at the NADH dehydrogenase complex (32). This is also confirmed by a complex perfused rat heart preparation where OHB increased contractility and decreased oxygen consumption (33). Furthermore it was shown, that ketogenesis increased parallel with growing brain size, which only uses glucose and KB as energy source, during the evolution of vertebrates (34). However, the highest part of KB's myocardial oxygen consumption can be found within fetal organisms (e.g. not over 7% in lamb) (35). Normal individual blood levels are less than 0.1 mM for both OHB and AAc, but can reach up to 8 mM of OHB and 2 mM for AAc and Acetone (36). Clinical Studies carried out by Lommi et al. showed that patients suffering from CHF do have a 2-fold increase in blood KB in mean compared to control subjects. Additionally blood KB levels correlate with severity of symptoms, degrees of venous congestion, left ventricular dysfunction as well as neurohormonal activation (9). Whereas previous studies assumed that KB block the citrate cycle (CC) (37) and thureupon cause contractility dysfunction (38) newer studies showed that KB prevent glucose uptake by inhibition of AMPK, generation of oxidative stress and promoting IR what might reduce cardiac energy substrate support in CHF (39). If the effect of KB is a negative one inhibition of carnitine palmitoyl transferase-1 (CPT-1) as most regulated step in ketogenesis may be a therapeutic possibility (40).

Oxidative Stress

Several mechanisms mentioned before produce reactive oxygen species (ROS), for example electron transport chain complex I (41) and mitochondrial uncoupling (42) lead to further mitochondrial damage and thus ROS production by oxidizing the plasma membrane, proteins and DNA (43). Various reports confirmed that ROS generation plays an important role in both AHF and CHF by decreasing mitochondrial functionality as for example shown in murine models (44).

Microcurrent Therapy

Mueller et al. discovered that cardiomyocytes cultured under electrical microcurrent application increase cell proliferation, change MMP and tissue inhibitors of metalloproteinase (TIMP) expression, modify collagen I to collagen III ratio and decrease expression of pro-inflammatory proteins (i.e. IL-6 and TNF-a) as well as the expression of growth factors (i.e. TGF-β) (45). Experiments in Wistar-Kyoto rats (WKY) and spontaneous hypertensive rats (SHR) showed that pericardial electrical microcurrent application up-regulates levels of MMP-2 and MMP-9, downregulates TIMP-3, reduces collagen III and reduces pro-inflammatory proteins (46). These results demonstrate that microcurrent application might be a promising method to increase heart function during CHF (47).

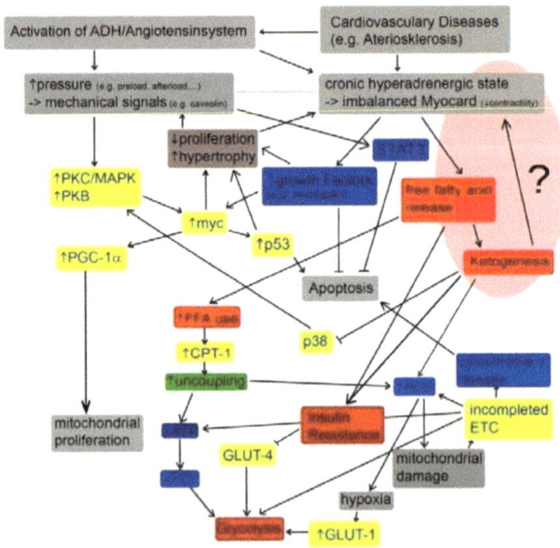

Figure 1: A schematic summary of the described mechanisms. The ellipse stands for the planed investigations. Starting with mechanical disturbance (e.g. high-blood-pressure as a result of chronic increased Angiotensin) various cell alterations result in increased glycolysis and fetal like metabolism. Growth factors inhibit apoptosis, however, the increased activation of myc results in hypertrophy leading to further mechanical myocardial imbalance, which is the vicious circle finally leading to heart failure.

Aim of this experiment

In this studies the effect of 3-hydroxybutyrate (OHB) on untreated and hypertrophic H9c2 cells was investigated. Hypertrophy could be induced by phenylephrine (PE) which binds to the β-adrenoreceptor and cross-activates caveolins, which further activates STAT3 and finally induces hypertrophic growth (4).

Since little is known about the effect of ketogenesis during the complex illness CHF it is interesting if OHB (and AAc – not in this experiments) improve, decrease or do not affect cardiomyocyte activity measured by using alamarBlue® and if they are a source of increased reactive oxygen species (ROS) generation. Additionally it was analyzed whether adrenergic stimulation and the supply of ketone bodies alter mitochondrial distribution and activity.

After establishment of a hypertrophic cardiomyocyte model including the effect of OHB on the cells the same experiments were performed by using primary cultures of cardiomyocytes isolated from spontaneous hypertensive rats (SH-R), which also develop hypertrophy over time and with primary cells obtained from SH-R treated with microcurrent.

Comparable experiments as described in this paper have been carried out by using P19 mesenchymal mouse cancer cells which can be differentiated into cardiomyocyte like cells in 1% DMSO; however their physiological properties were unsuitable to get conclusive results (data not shown)

Results

Hypertrophic Model

H9c2

Measurement of cellular dimensions showed, that cells treated with 62,5µM Phenylephrine (PE) have been with approximately 2.000 µm² twice as big as untreated cells cultured under normal cell culture conditions. (Figure 2) To confirm the mechanism (4) via Caveolin-3 and β-adrenergic stimulation immunohistochemical staining procedures against Caveolin-3 has been carried out (Figure 1).

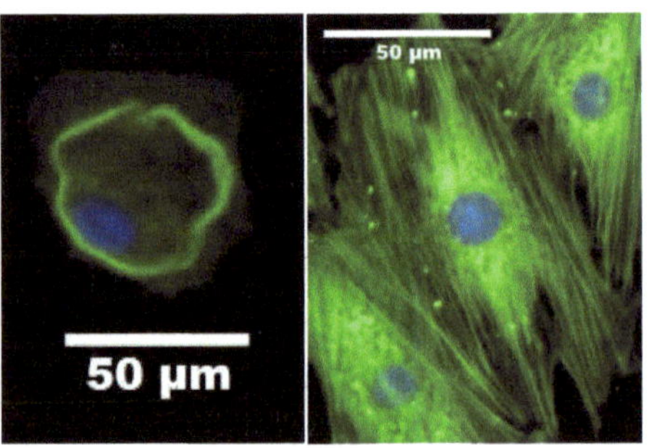

Picture 1 Immunohistochemical staining illustrates, that Caveolin-3 is located within the plasma membrane in untreated H9c2 cells (left), whereas in stimulated cells the movement into the cytoplasm can be seen clearly. (100x magnification)

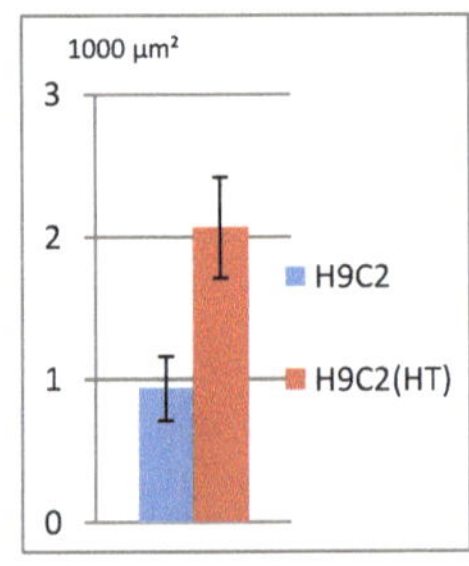

Diagram 1 Cells were measured with ImageJ as described in the attachment. With PE stimulated cells (H9c2(HT)) were about twice as large as untreated.

These results are corresponding to the observations of Jeong et al. and confirm the established model of hypertrophic H9c2 cells.

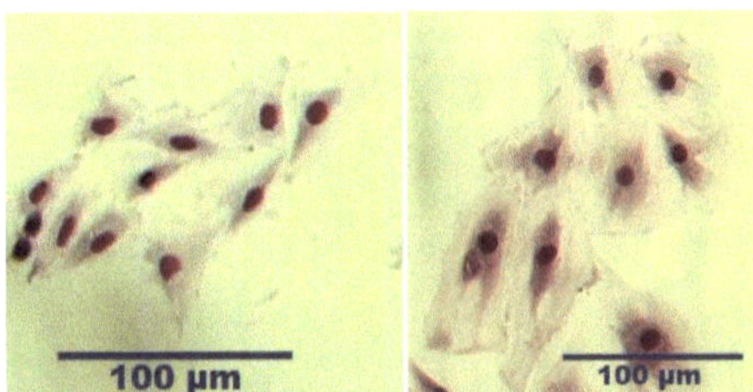

Picture 2 HE-Staining (carried out with Hemacolor Quick-Staining) of untreated (left) and with 50µM phenylephrine treated (right) H9c2 cells. The area of hypertrophic cells is twice as large compared to untreated cells.

Primary cardiomyocytes (SHR7)

SHR7 cells were isolated from a male 14 months old spontaneous hypertensive rat treated with microcurrent for 5 days. Immunohistochemical staining procedures against Caveolin-3 were performed similar to H9c2 cells. Untreated cells showed Caveolin-3 within the cytoplasm, as well as agglomeration at the cell membrane. With 62,5 µM phenylephrine stimulated cells showed caveolin-3 vesicles clearly within the cytoplasm. Interestingly treated cells as well as untreated SHR7 cells sporadically showed sickle-shaped caveolin-3 agglomeration at the plasma membrane.

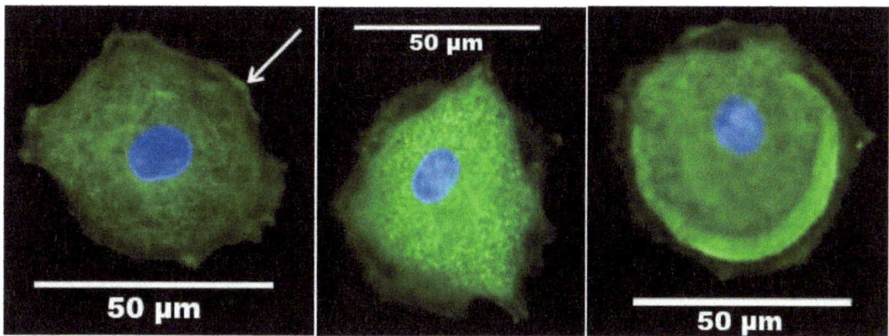

Picture 3 Immunohistochemical staining of SHR7 cells; untreated cells show Caveolin-3 within the cytosol and at the plasma membrane (left). PE stimulated cells clearly form cytoplasmic vesicles containing Caveolin-3 (middle). Some cells (both treated or not) show sickle-shaped Caveolin-3 at the plasma membrane (right).

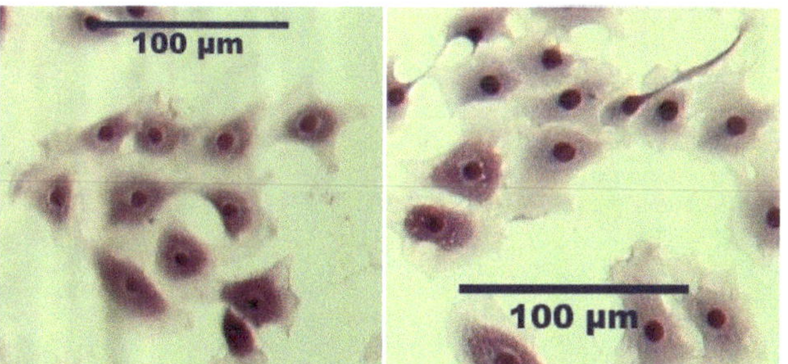

Picture 4 Hemacolor Quick staining of SHR7 cells untreated (left) and treated with 100µM phenylephrine (right). Both cell types appeared much hypertrophied. Exact measurement shows that PE treated cells were a quarter larger on average compared to untreated cells (see diagram 5).

Electrical Microcurrent treated primary (SHR7) cells

Electrical microcurrent treatment was achieved by electrodes placed into the medium and applied for 5 days. For negative control the same plate was used, however, without applying any current. HE-Staining showed that hypertrophic growth was diminished (Diagram 2). Caveolin-3 staining showed no difference compared to untreated cells. In both cases Caveolin-3 was located within the cytoplasm signaling hypertrophic growth.

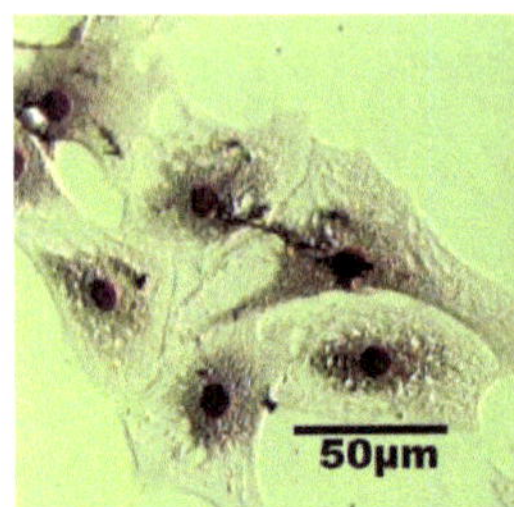
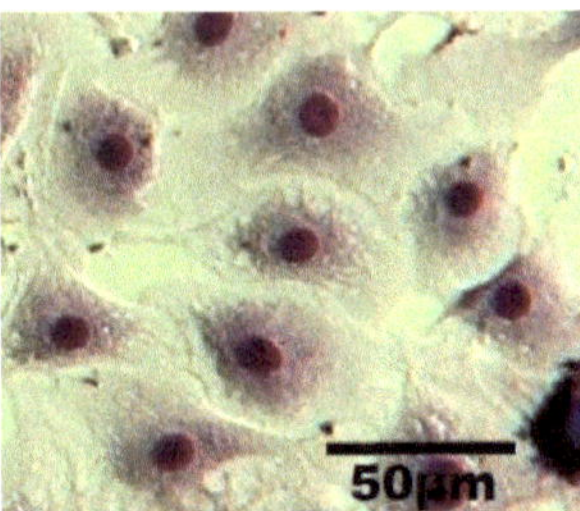

Picture 5 Hemacolor-Staining of cells treated with electrical microcurrent (right) and untreated (left). Cells were slightly smaller after microcurrent treatment.

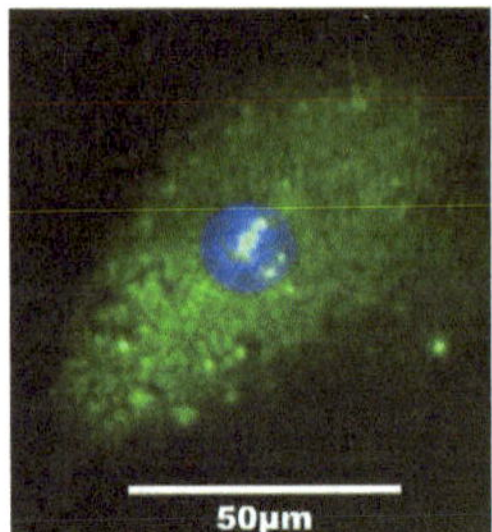
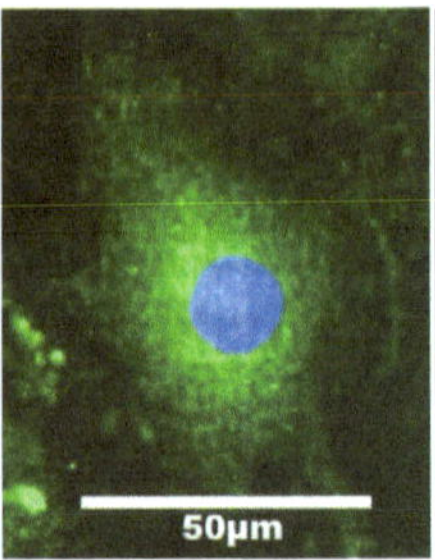
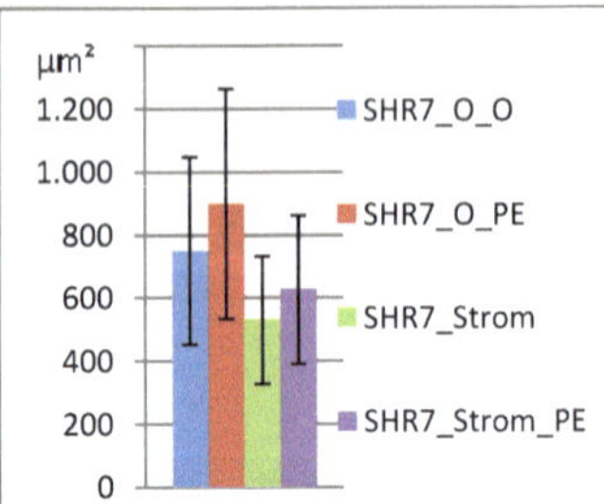

Picture 6 Caveolin-3 staining showed no difference in the location of Caveolin-3 in non-treated (left) and within microcurrent-treated (right) cells. In both cases Caveolin-3 was located in the cytoplasm

Diagram 2 Measurement of the cellular area stained by HE-staining of treated and untreated cells. Untreated SHR7 cells (750µm²+/-298) and cells stimulated by 50µM PE (898µm²+/-365) were larger compared to cells treated with electrical microcurrent (530µm²+/-203) and stimulated electrical microcurrent treated cells (627µm²+/-326)

Cell growth under β-Hydroxybutyrate treatment

Cell proliferation Assay with alamarBlue®

To define proliferation of normal and hypertrophied cells after treatment with concentrations from 200mM to 0.5mM of OHB an alamarBlue® assay was performed. Measurements were repeated four times by using H9c2 and three times with SHR7. Hypertrophied cells (treated with 100µM Phenylephrine) showed a statistically not significant increasing activity when treated with 3-12mM OHB compared to cells cultivated in 1% FCS containing culture medium. Cells cultivated in standard culture medium (D-MEM, 10%FCS, 1%PenStrep) showed no reaction based on OHB treatment.

In SHR7 cells no dose dependent reaction to OHB could be observed when cells were cultured in Cardiomyocyte-Medium (CM). Interestingly the cells needed much more time to turn over alamarBlue® (4h compared to 2h) to reach equal RFU comparable to H9c2. Dose-dependent reactions were observed in SHR7 cells cultured in serum-free medium by using concentrations of OHB between 2mM to 10mM. Higher or lower doses showed no difference in any cellular activity in CM as well as in SFM. Stimulation with PE resulted in no difference in the alamarBlue® assay.

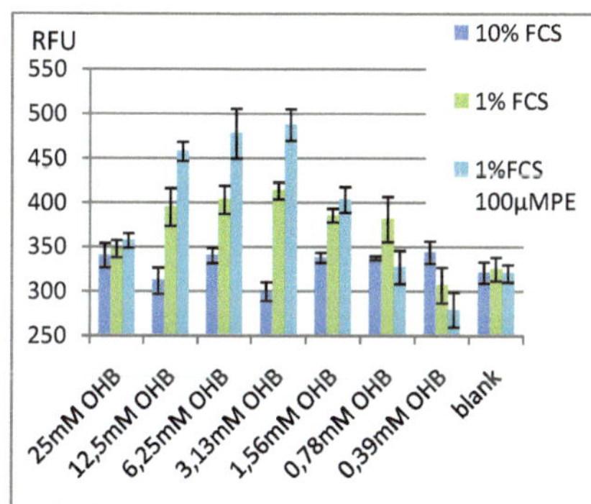

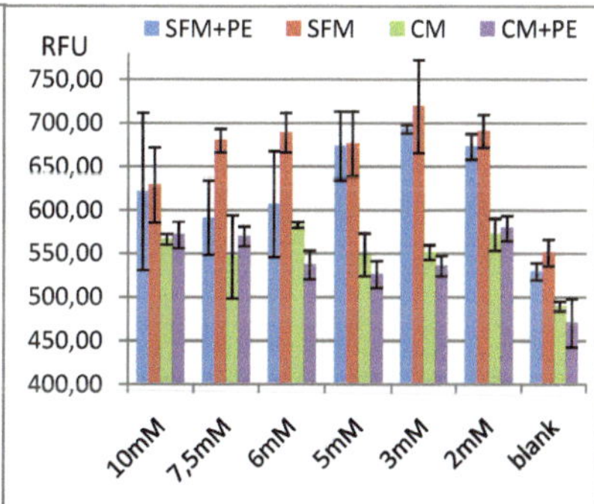

Diagram 2 Hypertrophied H9c2 cells (cyan) showed slightly increasing activity in the alamarBlue® assay compared to PE untreated cells (green). Controls showed no dose dependent reaction to OHB.

Diagram 3 SHR7 cells cultured in CM (green and violet) and SFM (blue and red) and stimulated with 50µM PE (blue and violet) and unstimulated (green and red). For cells cultured in CM no dose-dependent reaction was observed. Cells cultured in SFM showed increased activity within the range of 2 to 10mM OHB. PE showed no effect in the assay

Hypertrophic growth under β-Hydroxybutyrate treatment

Treatment with OHB showed that untreated cells were significantly larger than treated cells. The area of H9c2 cells and SHR7 cells stimulated with phenylephrine (H9c2: 2.012 µm² +/- 409,94; SHR7: 1.303 µm² +/- 361) could be reduced with 6,25mM OHB to the size of untreated cells (H9c2: 1031 µm² +/- 244; SHR7: 1061 µm² +/- 254). However, morphologically the cell shape differs compared to untreated cells (see picture 7 and 8). Hemacolor Quick staining pointed out a reduction of the extended cytoplasmic appendices clearly. Whereas untreated H9c2 cells appeared more elongated, after treated with OHB the cell shape seemed to be more round.

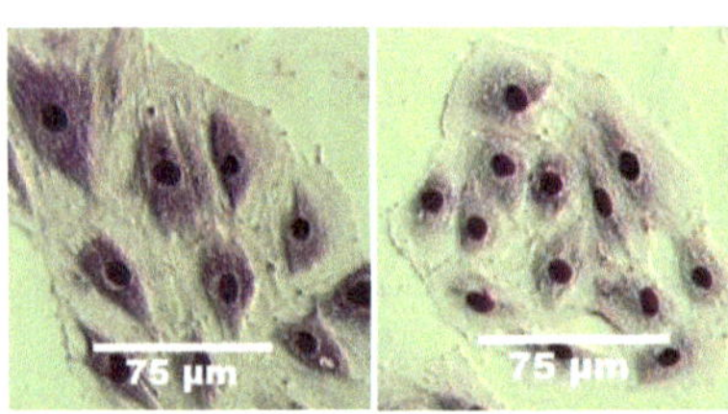

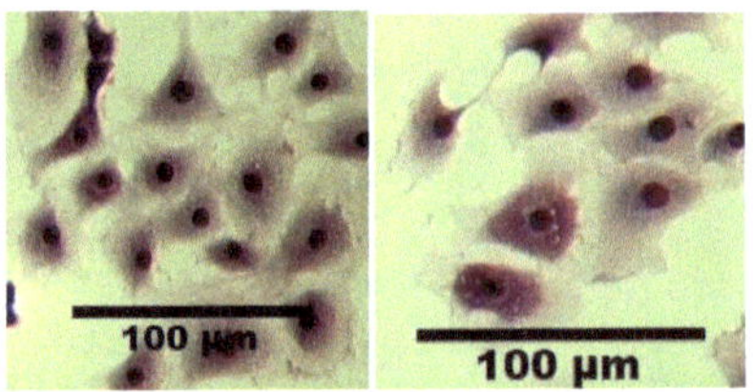

Picture 7 H9c2 cells treated with 62,5µM PE (left) and treated with 62,5µM PE and 6,25mM OHB (right). The areas of the latter one are significantly smaller then only stimulated with PE.

Picture 8 SHR7 cells treated with 125µM PE (left) and treated with 62,5µM PE and 6,25mM OHB (right). Cytoplasmic appendices were significantly smaller than PE stimulated cells.

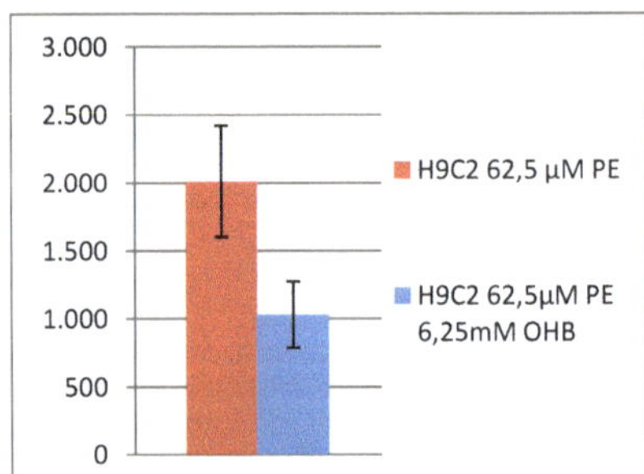

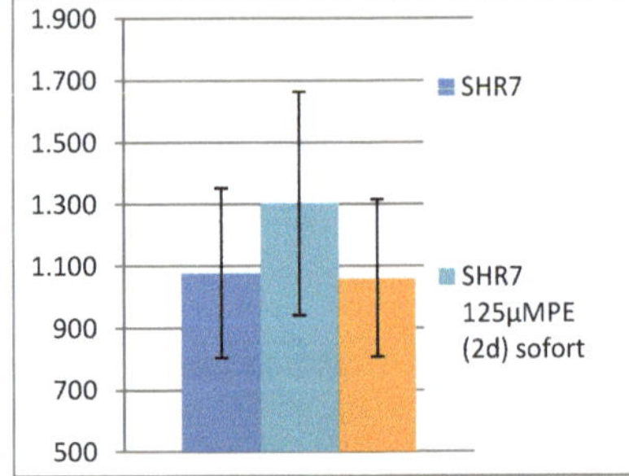

Diagram 4 Comparison of cell area of stimulated H9c2 (red) and stimulated H9c2 treated with 6,25mM OHB (blue). The area of treated cells was half as large, as for natural untreated cells.

Diagram 5 Comparison of cell area of unstimulated (blue), stimulated (cyan) and stimulated and treated with 6,25mM OHB (orange) SHR7 cells. OHB reduces the area of 2 days stimulated cells back to original cell size.

ROS assay

For measurement of β-hydroxybutyrate influence on ROS generation a ROS assay was
established. Cells were incubated with 10µM DCF (Invitrogen, Lofer, Austria) for 30 minutes
and then treated with 0.5mM to 200mM OHB as well as H_2O_2 for positive control. However in
multiple attempts results have not been reproducible (Data not shown). Further adoption of
the assay protocol or other methods would have to be necessary to obtain more reliable
results.

Immunohistochemical staining (Mitochondrial distribution)

H9c2

Mitochondrial staining by using MitoTracker® (Invitrogen, Paisley, UK) was performed to
visualize mitochondrial distribution within the cells including also changes during induced
hypertrophy and treatment with OHB. Untreated H9c2 cells appeared much brighter than
hypertrophied cells. Furthermore in hypertrophied cells only few mitochondria went into the
cytoplasmic appendices. Additionally the distribution of the mitochondria seemed to be more
granulated in the hypertrophied state and even more numerous when treated with OHB
compared to the original state. However, only stimulated cells do not seem to be distributed
throughout the cells equally after 72 hours of treatment contrary to hypertrophied cells treated
with OHB.

To verify the specifity of the mitochondrial staining and to show correlation of the distribution
with the density of the cytoskeleton, staining of mitochondria and immunohistochemical
staining of α-Tubulin has been performed. This staining confirmed the low density of
mitochondria in the extensive cytoplasmic appendices within the hypertrophied state.

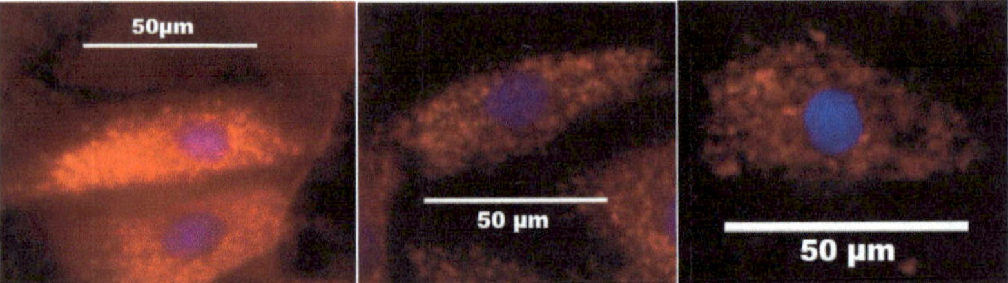

Picture 9 H9c2 cells untreated (left), treated with 62,5µM PE (middle) and treated with 62,5µM PE and 6,25mM OHB (right) stained with MitoTracker® and DAPI. Untreated cells show a far brighter signal, than hypertrophied cells. Furthermore the distribution of mitochondria seems to be more granular like when treated with OHB. However within only stimulated cells mitochondrial distribution seems to be less confluent compared to cells additionally treated with OHB.

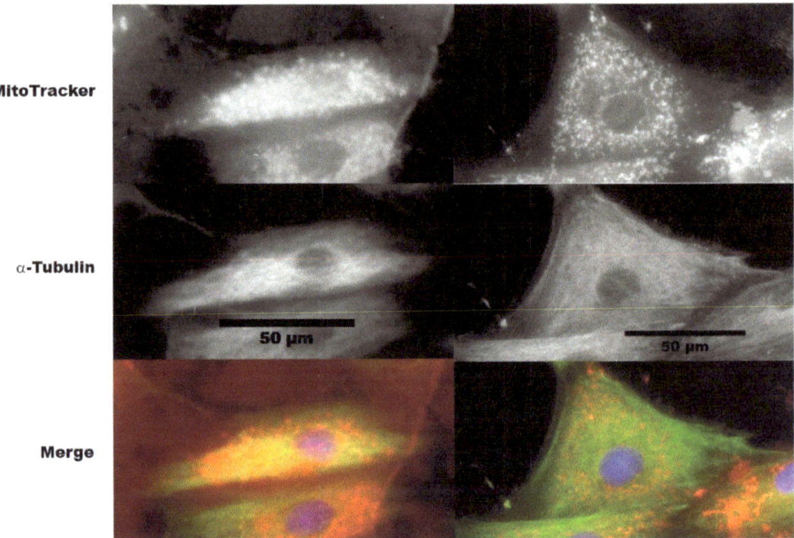

Picture 10 Untreated (left) and with 62,5µM PE hypertrophied cells (right) were stained with MitoTracker, anti-α-Tubulin (AlexaFlour 488) and DAPI. The signal of the MitoTracker (above) compared to the signal of α-Tubulin (middle) shows, that while within untreated cells mitochondrial distribution is confluent, the appendices of PE hypertrophied cells lack mitochondria.

SHR7

SHR7 cells showed a more loose mitochondrial distribution compared to H9c2 cells. Furthermore there was no difference in mitochondrial arrangements within SHR7 cells and microcurrent treated SHR7 cells. Stimulation with PE resulted in smaller but numerous mitochondria which also showed a denser intracellular distribution.

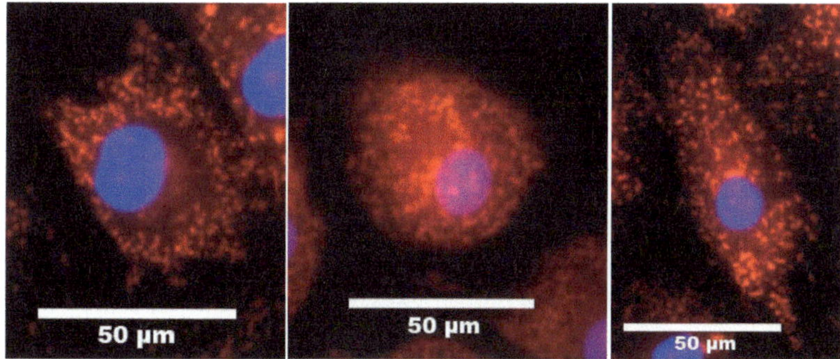

Picture 11 SHR7 cells stained with MitoTracker (400x magnification). Cells were left untreated (left) stimulated with 50µM PE (middle) and treated with microcurrent (right). Cells untreated and treated with microcurrent showed no difference in mitochondrial distribution. Compared to H9c2 mitochondria a more spread and less dense. PE treated cells showed smaller but more mitochondria packed together.

Discussion

Hypertrophic Model

H9c2 cells showed - as predicted and confirmed by immunostaining - hypertrophic growth as reaction to cellular stimulation with phenylephrine. Caveolin-3 entered the cellular cytoplasm and activated STAT3 and further proteins (4). Therefore the hypertrophic model has been successfully established. Compared to H9c2, primary cells (SHR7), which had already been treated with microcurrent in vivo before cell isolation, showed a nearly identical distribution of Caveolin-3 between the cell membrane and the cytoplasm. Internalization of Caveolin-3 could be reached clearly by stimulation with PE. Since primary cells without microcurrent treatment grow more slowly than SHR7 no investigation concerning the membrane localization of Caveolin-3 in native isolated SHR-cardiomyocytes were carried out within these studies. However in vitro application of microcurrent did not alter cytoplasmatic location of Caveolin-3 which would assume that reducing hypertrophy by microcurrent application is not reached by inhibiting adrenergic, or Caveolin-3/STAT3 signaling. Immunohistochemical staining of SHR hearts could show how much the Caveolin-3/STAT3 pathway is activated in vivo and uncover its importance in hypertrophy. In any case adrenergic stimulation does play a role in vertebrates with CHF and must not be neglected in the establishment of a cell culture model.

Mitochondrial Activity Assay

Elevated mitochondrial activity was reached by 3-hydroxybutyrate applied in concentrations within the range of 1mM to about 15mM when cells were cultured in media containing low or no serum. Cells cultured in high-serum concentrations (10%FCS) showed no dose-dependent reaction. This observation could indicate that OHB is an important energy source for cardiomyocytes starving or in poorly perfused tissue and its uptake seems not to be regulated by extracellular mechanisms. The results obtained from the alamarBlue® assay correlates with these observation which predicted that a maximum of energy can be obtained from OHB when concentrations range from 3 to 6mM (25) (36). However, it cannot be concluded that elevated OHB serum levels are of compensative nature, since also adverse effects of OHB by blocking glucose uptake have been described (39). Further investigations might uncover if elevated OHB serum levels previous observed in CHF patients (9) are compensative or only a symptom of the disease, but might therefore be of diagnostic value.

ROS Assay

Results received by performing ROS assays unfortunately were not reproducible. Several variations of testing protocol have been tried and lots of trouble shouting has been done, however the outcome was not satisfying. Pelletier et al. showed that concentrations of 5mM β-Hydroxybutyrate increases ROS production 2-fold (39). Although these results could not be confirmed it may be referred that oxidative stress induces insulin resistance (48), which could be treated by antioxidants (49) and increased glucose uptake during OHB treatment (39). Furthermore ROS can induce hypertrophy itself as recently shown (50) and must not be neglected during CHF. Therefore ROS generation of OHB has to be investigated in an in vivo model.

Immunohistocemical staining (Mitochondrial distribution)

Mitochondrial distribution shows slow adaptation to hypertrophic conditions in cells stimulated by phenylephrine. While mitochondria in untreated cells are distributed equally within the cytoplasm, in cells treated with PE some areas lacked mitochondria. Furthermore in adrenergic stimulated cells mitochondria appeared smaller and numerous compared to untreated cells which correlate with previous observations (11). Cellular MitoTracker-staining carried out after OHB treatment clearly showed a weaker fluorescent signal compared to treated cells. Since MT is metabolized by redox-reactions (similar to XTT) this might indicate mitochondrial damage (51). Cellular treatment using OHB after PE stimulation showed a better distribution of mitochondria within the cell which might be a sign for more ATP available for dynein and kinesin transporting mitochondria (52). This could indicated, that OHB provides a more efficient energy source in case of spontaneous hyperadrenergic stimulation, as in case of AHF orf CHF, and therefore allows faster adaptation and compensation for cardiomyocytes.

Conclusion

CHF is a very complex disease during which numerous variables play an important role. Adrenergic stimulation, blood ketone levels and many other factors influence the progression of this disease and therefore must be taken in account within a cell culture model. The role of ketone bodies in vivo remains still unclear and revealing its compensative or decompensative effect might be of great therapeutic value. However, bearing in mind that most variables influence each other isolated studies will not solve the problems leading to weaker cardiomyocytes, mechanical imbalance and finally decompensate heart failure and death. Microcurrent application could be a promising new therapeutic method as it showed a reduction of cellular hypertrophy in vitro as well as in vivo (47).

References

1. **Krüger, Wolfgang and Ludman, Andrew.** *Acute Heart Failure.* London : Birkhäuser, 2009.

2. **Venuta-Clapier, Renée and Garnier, Anne and Vladimir Veksler.** Energy metabolism in heart failure. *Journal of Physiology.* 2003, Vol. 555, pp. 1-13.

3. **Parton, Robert G. and Simons, Kai.** The multiple faces of caveolae. *Nature Reviews of Molecular Cell Biology.* 8, 2007, pp. 185-194.

4. **Jeong, Kyuho, et al.** Modulation of the caveolin-3 localization to caveolae and STAT3 to mitochondria by catecholamin-induced cardiac hypertrophy in H9c2 cardiomyoblasts. *Experimental and Molecular Medicine.* 2009, Vol. 41, pp.226-235.

5. **Chang, Mu-Hsin, et al.** IGF-II/mannose 6-phosphate receptor activation induces metalloproteinase-9 matrix activity and increases plasminogen activator expression in H9c2 cardiomyoblast cells. *Journal of Molecular Endocrinology.* 2008, Vol. 41, pp. 65-74.

6. **Lee, Hyoung-gon, et al.** Cell Cycle Re-Entry and Mitochondrial Defects in Myc-Mediated Hypertrophic Cardiomyopathy and Heart Failure. *PLoS ONE.* 4, 2009, Vol. 9, e 7172. doi:10.1371/journal.pone.0007172.

7. **Wikipedia.** Atrial natriuretic peptide. [Online] [Cited: 1 12, 2010.] http://en.wikipedia.org/wiki/atrial_natriuretic _peptide.

8. **Mettauer, B., et al.** Heart failure: a model of cardiac and skeletal muscle energetic failure. *European Journal of Physiology.* 2006, Vol. 452, pp. 653-666.

9. **Lommi, Jyri, et al.** Blood Ketone Bodies in Congestive Heart Failure. *Journal of the American College of Cardiology.* 1996, Vol. 28, pp. 665-672.

10. **Lee, Hsin-Chen and Wei, Yau-Huei.** Mitochondrial biogenesis an mitochondrial DNA maintenance of mammalian cells under oxidative stress. *International Journal of Biochemistry and Cell Biology.* 2005, Vol. 37, pp. 822-834.

11. **Sharov, Victor G., et al.** Abnormal Mitochondrial Function in Myocardium of Dogs with Chronic Heart Failure. *Journal of Molecular Cell Cardiology.* 1998, Vol. 30, pp. 1757-1762.

12. **Marín-García, José, Goldenthal, Michael and Moe, Gordon.** Abnormal cardiac and skeletal muscle mitochondrial function in pacing-induced cardiac failure. *Cardiovascular Research.* 2001, Vol. 52, pp. 103-110.

13. **Ide, Tomomi, et al.** Mitochondrial DNA Damage and Dysfunction Associated With Oxidative Stress in Failing Hearts After Myocardial Infarction. *Circulation Research.* 2001, Vol. 88, pp. 529-535.

14. **Williams, RS.** Mitochondrial gene expression in mammalian striated muscle. Evidence that variation in gene dosage is the major regulatory event. *J Biol Chem.* 1986, Vol. 261, pp. 12390-12394.

15. **Scheubel, Robert J., et al.** Dysfunction of Mitochondrial Respiratory Chain Complex I in Human Failing Myocardium Is Not Due to Disturbed Mitochondrial Gene Expression. *Journal of the American College of Cardiology.* 40, 2002, Vol. 12, pp. 2174-2181.

16. **Taha, Maysa and Lopaschuk, Gary D.** Alterations in energy metabolism in cardiomyopathies. *Annals of Medicine.* 2007, Vol. 39, pp. 594-607.

17. **Neubauer, Stefan, et al.** Myocardial Phosphocreatine-to-ATP Ratio Is a Predictor of Mortality in Patients With Dilated Cardiomyopathy. *Circulation.* 1997, Vol. 96, pp. 2190-2196.

18. **Taegtmeyer, Heinrich.** Switching Metabolic Genes to Build a Better Heart. *Curr Probl Cardiol.* 1994, Vol. 19, pp.59-113.

19. **Bilsen, Marc van, et al.** Metabolic remodelling of the failing heart: the cardiac "burn-out" syndrome? *Cardiovascular Research.* 2004, Vol. 61, pp. 218-226.

20. **Liao, Ronglih, et al.** Cardiac-Specific Overexpression of GLUT1 Prevents the Development of Heart Failure Attributable to Pressure overload in Mice. *Circulation.* 2002, Vol. 106, pp. 2125-2131.

21. **Belfort, R, et al.** Dose-response effect of elevated plasma free fatty acid on insulin signaling. *Diabetes.* 2005, Vol. 54, pp. 1640-1648.

22. **Marangou, AG, et al.** Hormonal effects of norepinephrine on acute glucose disposal in humans: a minimal model analysis. *Metabolism.* 1988, Vol. 37, pp. 885-891.

23. **Dulloo, A.G., Samec, S. and Seydoux, J.** Uncoupeling protein 3 and fatty acid metabolism. *Biochemical Society Transactions.* 2001, Vol. 29, pp. 785-791.

24. **Hickerson-Bick, Diane L.M., Jones, Chad and Buja, L. Maximilian.** Stimulation of mitochondrial biogenesis and autophagy by lipopolysaccharide in the neonatal rat cardiomyocyte protects against programmed cell death. *Journal of Molecular and Cellular Cardiology.* 2008, Vol. 44.

25. **Ashrafian, Houman, Frenneaux, Michael P. and Opie, Lionel H.** Metabolic Mechanisms in Heart Failure. *Circulation.* 2007, Vol. 116, pp. 434-448.

26. **Sasaoka, T, Wada, T and Tsuneki, H.** Lipid phosphatases as a possible. *Pharmacol Ther.* 2006, Vol. 112, pp. 799-809.

27. **Taylor, M, et al.** An evaluation of myocardial fatty acid and glucose uptake using PET with [18F]fluoro-6-thia-heptadecanoic acid and [18F]FDG in patients with congestive heart failure. *J Nucl Med.* 2001, Vol. 42, pp. 55-62.

28. **Witteles, RM, et al.** Insulin resistance in idiopathic dilated cardiomyopathy; a possible etiologic link. *J Am Coll Cardiol.* 2004, Vol. 44, 78-81.

29. **Santomauro, AT, et al.** Overnight lowering of free fatty acids with Acipimox improves insulin resistance and glucose tolerance in obese diabetic and nondiabetic subjects. *Diabetes.* 1999, Vol. 48, pp. 1836-1841.

30. **Opie, LH.** The metabolic vicious cycle in heart failure. *Lancet.* 2006, Vol. 364, pp. 1733-1734.

31. **Lommi, J, et al.** Heart failure ketosis. *J Intern Med.* 1997, Vol. 242, pp. 231-238.

32. **Veech, RL, et al.** Ketone bodies, potential therapeutic uses. *IUBMB Life.* 2001, Vol. 51, pp. 241-247.

33. **Kashiwaya, Y, et al.** Control of glucose utilization in working perfused rat heart. *J Biol Chem.* 1994, Vol. 269, pp. 20321-20334.

34. **Cahill, George F. and Veech, Richard L.** Ketoacids? Good Medicine? *Transactions of the American Clinical and Climatological Association.* 2003, Vol. 114, pp. 149-163.

35. **Bartelds, B., van der Leij, F.R. and Kuipers, J.R.G.** Role of Ketone Bodies in Perinatal Myocardial Energy Metabolism. *Biochemical Society Transactions.* 2001, Vol. 29, pp. 325-330.

36. **Cahill, GF Jr.** Starvation in man. *New Eng J Med.* 1970, Vol. 282, pp. 668-675.

37. **Russell III, Raymond R. and Taegtmeyer, Heinrich.** Coenzyme A Sequestration in Rat Hearts Oxidizing Ketone Bodies. *Journal of Clinical Investigation.* 1992, Vol. 89, pp. 968-973.

38. **Taegtmeyer, Heinrich.** Energy metabolism of the heart: from basic concepts to clinical applications. *Curr Probl Cardiol.* 1994, Vol. 19, pp. 59-113.

39. **Pelletier, Amélie and Coderre, Lise.** Ketone bodies alter dinitrophenol-induced glucose uptake through AMPK inhibition and oxidative stress generation in adult cardiomyocytes. *Am J Physiol Endocrinol Metab.* 2007, Vol. 292, pp. 1325-1332.

40. **Fukao, T, Lopaschuk, GD and Mitchell, GA.** Oathways and control of ketone body metabolism: on the fringe of lipid biochemistry. *Prostaglandins Leukot Essent Fatty Acids.* 2004, Vol. 70, pp. 243-251.

41. **Ide, Tomomi, et al.** Mitochondrial electron transport complex I is a potential source of oxygen free radicals in the failing myocardium. *Circulation Research.* 1999, Vol. 85, pp. 357-363.

42. **Saavedra, Walter F., et al.** Imbalance Between Xanthine Oxidase and Nitric Oxide Synthase Signaling Pathways Underlies Mechanoenergetic Uncoupling in the Failing Heart. *Circulation Research.* 2002, Vol. 90, pp. 297-304.

43. **Tsuitsui, Hiroyuki.** Mitochondrial Oxidative Stress and Heart Failure. *Internal Medicine.* 2006, Vol. 13, pp. 809-813.

44. **Ide, Tomomi, et al.** Direct Evidence for Increased Hydroxyl Radicals Originating From Superoxide in the Failing Myocardium. *Circulation Research.* 2000, Vol. 86, pp. 152-157.

45. *Normalisation Of The Cardiac Extracellular Matrix By Microcurrent Application With The Objective Of Healing Failing Hearts.* **Mueller J, Kapeller B, Losert UM, Macfelda K.** Umea : Talk at ESAO 2006, 2006.

46. **Macfelda, Karin, et al.** *Does the Application of Electrical Microcurrent Heal Heart Failure by Downregulation of the Pro-inflammatory Cytokines? First Pre-Clinical in Vivo Results.* 2009.

47. *Does the Direct Myocardial Application of Electrical Microcurrent Heal Heart Failure by Downregulation of the pro-inflammatory cytokines ?* **Mueller J, Kapeller B, Heinze H, Hofmann M, Hohlfeld J, Losert UM, Macfelda K.** s.l. : Poster Presentation ISHLT Paris, 2009.

48. **Evans, JL, Maddux, BA and Goldfine, ID.** The molecular basis for oxidative stress-induced inxulin resistances. *Antioxid Redox Signal.* 2005, Vol. 7, pp. 1040-1052.

49. **Haber, CA, et al.** N-acetylcysteine and taurine prevent hyperglycemia-induced insulin resistance in vivo: possible role of oxidative stress. *Am J Physiol Endocrinol Metab.* 2003, Vol. 285, E744-E753.

50. **Oyama, Kyohei, Takahashi, Kiyoshi and Sakurai, Koichi.** *Cardiomyocyte H9c2 Cells Exhibit differential sensitivity to Intracellular Reactiv Oxygen Species Generation with Regard to their Hypertrophic vs. Death Responses to Exogenously Added Hydrogen Peroxide.* 45 : s.n., 2009. pp. 361-369.

51. **Invitrogen, Molecular Probes.** MitoTracker Mitochondrion-Selective Probes. *Data Sheet.* 2008.

52. **Frederick, Rebecca L. and Shaw, Janet M.** Moving Mitochondria: Establishing Distribution of an Essential Organelle. *Traffic.* 8, 2007, pp. 1668-1675.

53. **Collection, American Type Culture.** *Cell Line Designation: H9c2(1-1) CRL-1446 (Data Sheet).* s.l. : ATCC, 1995.

Attachment:

Abbreviations

AAc	Acetoactate
AHF	Acute Heart Failure
CHF	Chronic Heart Failure
CM	Cardiomyocyte Medium
CO	Cardiac Output
DAPI	4',6'-Diamidin-2'phenylindol
DMEM	Dulbecco's Modified Eagel's Medium
ECM	Extracellular Matrix
FFA	Free Fatty Acids
HF	Heart Failure
IR	Insulin Resistance
KB	Ketone Bodies
MMP	MatrixMetalloProteinase
MT	MitoTracker®
OHB	β-Hydroxybutyrate
PCr	Phosphocreatine
ROS	Reactive Oxygen Species
SFM	Serum-Free Medium
SHR	Spontaneous Hypertensive Rat
SV	Stroke Volume

Materials

H9c2 cells were obtained from American Type Culture Collection (ATCC, Manassas, VA, USA) and are a subclone derived from embryonic BD1X rat heart tissue (53).

Cells were cultured in Dulbecco's Modified Eagle's Medium (DMEM) (Sigma-Aldrich, Steinheim, Germany) supplemented with 10% Fetal Calf Serum (PAA Laboratories GmbH, Pasching Austria) and 1% Penicillin/Streptomycin (Gibco, Invitrogen, Auckland, New Zealand); For subcultivation the cells were removed by using 5% Trypsin-EDTA (Invitrogen, Paisley, UK)

Sodium 3-hydroxybutyrate, R-Phenylephrinehydrochloride, Triton X-100 and DAPI were obtained from Sigma Aldrich (St. Louis, USA)

Reactive Oxygen Species were measured by Reactive Oxygen Species (ROS) Detection reagents [2',7'-dichloroflurescein (DCF)] (Molecular Probes, Invitrogen, Lofer, Austria)

AlamarBlue® Assays were performed by using alamarBlue® Cell Viability Reagent, for MitoTracker® images MitoTracker® Mitochondrion-Selective Probes was used (both Molecular Probes, Invitrogen, Paisley, UK)

The Caveolin-3 antibodies was purchased from Santa Cruze Biotechnology, Inc. (Santa Cruz, USA), the α-Tubulin antibodies were obtained from Sigma-Aldrich (St. Louis, USA) and Alexa Fluor 488 goat anti-mouse IgG from Molecular Probes (Invitrogen, Paisley, UK)

For washing steps Dulbecco's Phosphate Buffered Saline (Sigma-Aldrich, Steinheim, Germany) was used.

Slides were fixed with Dako Fluorescent Mounting Medium (DakoCytomation, Inc, Carpinteria, California, USA).

Fluorescence images were taken by using Nuance multispectral imaging system (Cambridge Research & Instrumentation, Inc., Cambridge UK) installed on an Olympus BX60 Microscope (Olympus Europe, Hamburg, Germany). The used objectives were Olympus UplanFl 20x/0,50 8/0,17 and Olympus UplanFl 40x/0,75 8/0,17.

Brightfieldimages were taken by the use of a Nikon Digital Net Camera DN100 (Nikon Instruments Inc., New York, USA) installed on an Olympus IMT-2 inverse Microscope with an Olympus 10x/0,30 160/0,17 objective. (Olympus Europe, Hamburg, Germany)

Photometric and fluorescence measurement was carried out by using a Varioskan Flash (Thermo Electron Corporation, Thermo Fisher Scientific, Waltham, Massachusetts)

Hematoxilin/Eosin-Staining was performed by using Hemacolor Quick Staining Solutions obtained from Merck (Darmstadt, Germany). Cristal violet also obtained from Merck (Darmstadt, Germany) was used in a concentration of 0,2%w/v solved in 2%v/v Ethanol.

SDS was obtained from Sigma (Steinheim, Germany).

SHR7 cells were isolated mechanically from a cardiac biopsy of a 14 month old male SHR which has already been treated with microcurrent for 5 days. Cells were cultured in a specific culture medium (cardiomyocyte-medium [CM]) containing D-MEM, 10% FCS, 0,1 mM NEAA, 1% PenStrep, 1% ascorbic acid, 10ng/ml insulin, 10ng/ml transferrin and 20ng/ml sodium selenite.

Methods

Hypertrophic Model

According to the hypertrophic model established by Jeong et al. (4) H9c2 cells were incubated with DMEM (Sigma-Aldrich) and 10% FCS (PAA Laboratories) for 24h. The cells were starved in DMEM with 1% FCS for 24h and then incubated with 50µM to 100µM Phenylephrine (PE) (Sigma) for 48h.

Hypertrophic Measurement

Cells were treated in a 6-well cell culture plate. From each well 5 various pictures were taken. 200 cells were calibrated by using ImageJ software.

Hematoxilin/Eosin-Staining

Cells were fixed and stained with Hemacolor quick staining (Merck) and treating the cells with solution 1 (methanol), 2 (reagent red) and 3 (reagent blue) for 30 seconds each.

MitoTracker®

Cells were washed with PBS (Sigma) and incubated with DMEM (Sigma-Aldrich) containing 300nM MT for 4h. Slides were washed with PBS and fixed with warm formaldehyde for 15 minutes. Cells were fixed with ice-cold acetone for 10 minutes. A counterstaining was carried out using 1µg/ml DAPI for 5 minutes. Slides were covered with Dako Fluorescence Mounting Medium and imaged with Nuance.

ROS-Measurement

Cells were washed with PBS and loaded with DMEM supplemented with 10µM DCF (Molecular Probes) for 30 Minutes. Thereafter treatment with OHB and H_2O_2 (as positive control) and catalase (as negative control) was carried out. Cells were lysed with Triton X-100 (Sigma) and fluorescence was measured by the use of a Varioskan Flash reader (Thermo).

OHB Proliferation Assay

To test whether OHB promotes cell viability within a hypertrophic state H9c2 cells were treated for simulating hypertrophy or left untreated by carrying out only a change of culture medium as described before. SHR7 cells were left in specific cell culture medium for 24h. Then all cells were incubated with the respective concentration of OHB for 48h. Thereafter an alamarBlue® Assay was carried out. Tests were repeated at least 3 times. Each value was determined 3-fold per test.

AlamarBlue®-Assay

Adherent cells cultured and treated in a 96-well cell culture plate were incubated with 20µl alamarBlue® (Invitrogen) per well at 37°C and 5%CO_2 for 2h for H9c2 cells and for 4h for SHR7 cells. Fluorescence was measured by using a Varioskan Flash reader (Thermo) [Extinction 570nm; Emission 585nm].

Cristal Violet-Assay

Plates were fixed with Cristal Violet for 15 minutes, washed with H_2O 3 times and dried overnight. Cells were lysed with 1%w/v SDS and photometric measurement was done with Varioskan Flash reader.

Immunohistochemical staining (Caveolin-3)

Slides covered with untreated, hypertrophied or OHB-treated cells were fixed with ice-cold acetone for 10 minutes. The cells were cracked by using 3% H_2O_2 for 5 minutes, thereafter washed with H_2O and PBS for 5 minutes. After a blocking step with 10% goat-serum for 20 minutes and washing with PBS for 2 minutes, cells were incubated with primary antibodies at 4°C overnight. Cells were incubated with the secondary antibody (goat-anti-mouse conjugated with AlexaFlour488) 1h at room temperature and slides were thereafter washed with PBS for 5 minutes. Cellular counter-staining was performed by using DAPI for 5 minutes and slides were mounted with Dako Fluorescence Mounting Medium finally.

Immunohistochemical staining (α-Tubulin)

Slides covered with H9c2 cells treated with MitoTracker® as described before were fixed with 37°C warm PFA for 15 minutes. Cells were washed by using ice-cold Aceton for 10 minutes, dried and washed with TBS for 5 minutes. Cells were cracked with 3% H_2O_2 for 10 minutes. After blocking with 10% serum for 20 minutes, cells were incubated with the primary antibody overnight at 4°C, thereafter with the secondary antibody 1h at room temperature. Cellular counter-staining was carried out with DAPI for 5 minutes and finally th slides were mounted with Dako Fluorescence Mounting Medium.

Microcurrent treatment

SHR7 as well as H9c2 cells were seeded on cover slips put into a modified 24-well cell culture plate, which includes a pair of electrodes per well. The whole cell culture plate was connected to a specific electronic device (see picture 11) and cellular microcurrent treatment was carried out for 5 days (45).

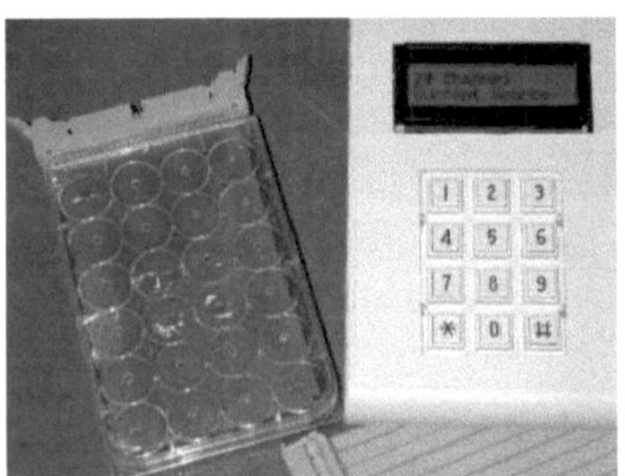

Picture 12 the electronic device

Thanks

To all my colleagues especially to Dr. Karin Macfelda for supervising and to Barbara Kapeler for her professional advices.

A special thanks to Roman Lieber, who took all nice fluorescence images for me.

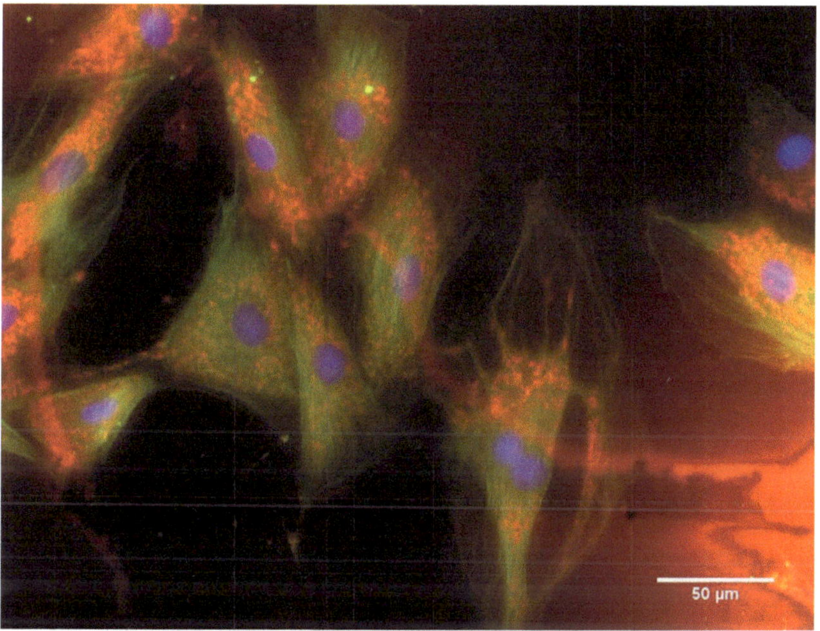

Picture 13 **Fluorescence Image of H9c2(HT) cells treated with 62,5 µM Phenylephrine. Blue: Nucleus (DAPI), Red: Mitochondria (MitoTracker®), Green: microtubule (α-Tubulin).**